Nelson Interı Science

Workbook 3

Nelson Thornes

Published in 2012 by:
Nelson Thornes Ltd
Delta Place
27 Bath Road
CHELTENHAM
GL53 7TH
United Kingdom

12 13 14 15 16 / 10 9 8 7 6 5 4 3 2 1

A catalogue record for this book is available from the British Library

ISBN 978 1 4085 1728 4

Cover illustration: Andy Peters
Illustrations by Maurizio de Angelis, David Benham, Simon Rumble and Wearset Ltd
Page make-up by Wearset Ltd, Boldon, Tyne and Wear
Printed by Multivista Global Ltd

Acknowledgements

The authors and the publisher would like to thank Judith Amery for her contribution to the development of this book.

Contents

Contents

Introduction

Nelson International Science Workbook 3 provides a complete copy of the *Student Book* activities for all learners to work through.

The activities are marked with 📖 showing the corresponding page number in the *Student Book*.

In addition to the *Student Book* activities, there are extra activities marked, for example, Activity A, that can be done in the classroom or as homework at home. They support the knowledge and understanding gained in the *Student Book* activities.

Plant parts

Every plant has parts. Look at the picture. It shows some plant parts. Can you identify them?

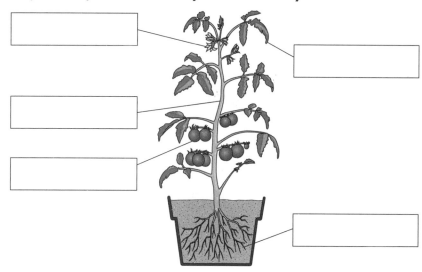

Activity 1

You will need: a pen or pencil.

 1 Look at the picture of the tomato plant above.

 2 Label it with the names of the plant parts.

Here are the names with the letters mixed up. You will have to sort them out:

t o r o _____

f a l e _____

u r t i f _____

s m e t _____

w f o l r e _____

3 Show your work to the class.

Activity 2: Compare two plants 3

You will need: access to some plants and a pen or pencil.

1 At school, go outside and find two different plants. Try to find ones with flowers.

2 Please check with your teacher before you dig up any plants. If you can dig them up, then do it carefully and take the plants back to class.

a Observe them carefully.
Make drawings of them on the next page.

b Write the names of the parts on the drawings too.

Look at the examples shown here. These are just some of the plant parts that you can label on your drawings.

> petal   stem   sepal   flower
> root   leaf   fruit

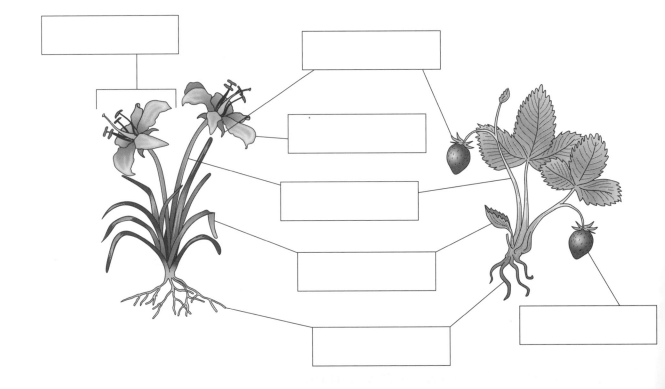

Activity 2: Compare two plants (continued) 📖 4

Draw and label your two plants here.

Plant 1   **Plant 2**

3 If you cannot dig up two plants where they are growing, make drawings of them. Label the parts you can see.

4 Compare the two plants and compare your two drawings.

a Look for similarities and differences.

b Make lists of the similarities and differences.

Similarities	Differences

5 Display your drawings and lists for the class to see. Look at what others have done.

What plants need to grow: water and light

Activity 3 5

You will need: some seedlings and a pen or pencil.

1 What evidence could you collect to show that plants need water and light to grow well?

 a In school, discuss ideas with your group. Try to plan a f[...]
test.

b Write down the group's plan and show it to your teache[...]

Plan
First we
Then we
Next we
The test will be fair because

c Collect the things you will need. List them here:

Activity 3 (continued) |5|–|6|

2 Decide what evidence you will collect and how you will record it. If you want to use a table or a bar chart, ask your teacher for help.

Record your evidence here:

3 Write down your own predictions about what will happen.

Prediction
I predict that _____
_____
_____
_____

4 Share your plan and predictions with the class, and listen to the other groups.

a Compare their ideas with yours.

b Do you need to change anything about your test before you start?

5 Set up your test.

Activity 3 *(continued)* 7

 6 Drawings or measurements are suitable ways of recording change.

a Make a record of your observations at the time the test begins.

Draw the pots every week for four weeks.

b Record the time and date of your observations.

Make your drawings on the same day each week.

Week 1

Date: _____

Time: _____

Week 2

Date: _____

Time: _____

Week 3

Date: _____

Time: _____

Week 4

Date: _____

Time: _____

7 Make regular observations and record them.
Continue until you can see no more change.

8 Compare your results with your predictions.

a Were the predictions correct? Tick the correct box.

Yes ☐   No ☐

b Look at your results and use them to make conclusions about the idea you were testing.

c Write down the group's conclusions.

_____

_____

_____

_____

9 In class, present your results and conclusions in a display.

a Look at the other groups' results. Compare them with yours.

b Explain your results and conclusions and ask other groups to explain theirs.

How water is taken in and transported

Activity 4: Exploring the function of roots 📖 10

You will need: a small plant, water, a container, a plastic bag, string and a pen or pencil.

1 Put water into the container and fill it to the top. Carefully dig up a small plant, taking care not to break off its roots.

2 Lower the plant into the container and allow water to overflow.

a Wrap the plastic bag around the container.

b Use the string to close the bag tightly around the stem of the plant. This should hold the stem above the water level.

Activity 4: Exploring the function of roots *(continued)* 📖

3 Use the table below to record the water level each day, over at least a week.

Day	Height of water (mm)
Monday	
Tuesday	
Wednesday	
Thursday	
Friday	

a Measure the height of the water in the container in millimetres (mm).

b Record the height in the table.

 4 Repeat the measurement and recording each day.

Activity 4: Exploring the function of roots *(continued)*

5 When the period of measuring is over, use the measurements you have recorded to make a bar chart showing what happened to the water level.

The horizontal axis should show the days of the week, and the vertical axis should show the level of the water.

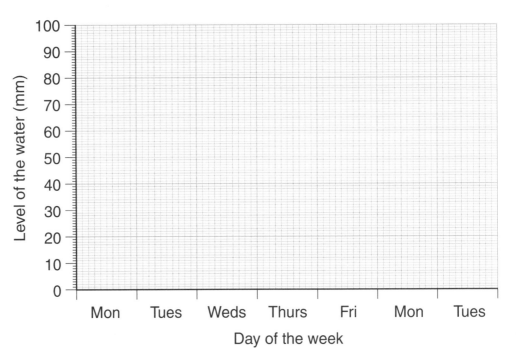

6 Write a report of what you did in the investigation here:

Report

First we

Then we

Next we

Activity 4: Exploring the function of roots (continued) |12|

7 Look at your bar chart. What does it show about the function of the roots?

a Write down your ideas.

_____

_____

_____

_____

b Share your conclusions with the class.

c Discuss the conclusions that different learners have.

8 Try to answer these questions:

a Why was the top of the container covered with a plastic bag and closed around the stem?

_____

_____

b Why did the level of the water change?

_____

_____

c What would have happened if the roots had been cut off before the plant was put in the container?

_____

_____

_____

_____

Healthy roots, stems and leaves

Activity 5 📖 13

You will need: a pen or pencil.

Plant A   **Plant B**

1 Look at the two plants in the picture.

2 Compare them and make a list of the differences you can see.

Plant A

_____

_____

_____

Plant B

_____

_____

_____

3 Discuss your list with others in your group.

4 Try to explain the differences you have observed.

_____

_____

_____

_____

Activity 6: Investigate if plant growth is affected by temperature 16

You will need: a pen or pencil.

1 In school, discuss with your group how to test this.

2 Write down the question you will try to answer.

_____

_____

3 Make a plan for a fair test of your idea.

a Change only one factor and keep all the others unchanged.

b Show the plan to your teacher.

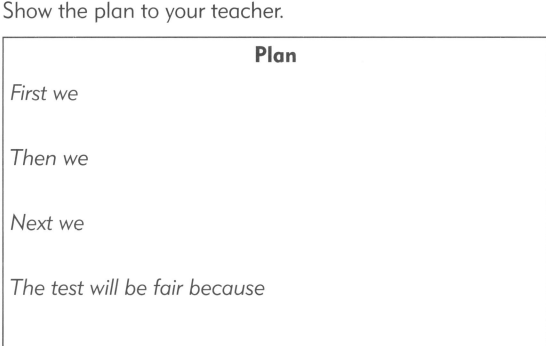

Plan

First we

Then we

Next we

The test will be fair because

Activity 6: Investigate if plant growth is affected by temperature *(continued)* 16

4 Write down your prediction of what you expect to happen.

I predict that _____

_____

_____

_____

5 Collect the things you need. List them here:

Decide what you will observe and how you will record your observations.

6 Carry out the test until you have an answer to your question.

7 What did you conclude from your results?

a Write down your conclusions.

I found out that _____

_____

_____

b Compare the results with your prediction. What was the difference, if any?

_____

_____

_____

8 In class, present your results and conclusions. Try to explain what you have found.

Activity A

You will need: a pen or pencil.

Complete these sentences using these words (you may need to use some words more than once):

> faster   plants   stem   higher   leaves
> grow   temperature   plant   roots
> stems   light   water

1. Plants have leaves, __ __ __ __ __, __ __ __ __ __ and flowers.

2. Water and __ __ __ __ __ are needed by __ __ __ __ __ __ for them to __ __ __ __.

3. The __ __ __ __ __ take __ __ __ __ __ from the soil and it travels through the __ __ __ __ to all parts of the __ __ __ __ __.

4. If the roots, __ __ __ __ __ __ __ and __ __ __ __ are not healthy, the plant will not __ __ __ __well.

5. The __ __ __ __ __ __ __ __ __ __ __ __ in greenhouses can be kept __ __ __ __ __ __ and this makes plants grow __ __ __ __ __ __.

Activity B

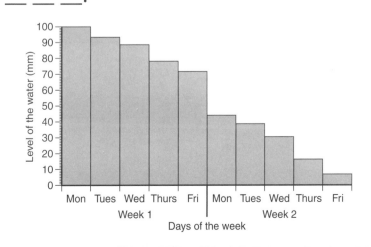

Activity B (continued)

You will need: a pen or pencil.

1 The bar chart on page 17 shows the results of an investigation of water uptake by the roots of a plant (as in Activity 4, page 10).

2 Look at the bar chart and answer these questions.

a Why does the level of the water go down?

_____

_____

b Why is the change in water level not the same every day?

_____

_____

c When is the change in level greatest?

_____

Why?

_____

d What is the total amount of water lost over the two weeks?

_____

_____

e Which two days had the smallest change of level?

_____

Try to explain why.

_____

_____

Life processes

Here are some life processes common to all
animals and people.

Activity 1 ⟦18⟧

You will need: a pen or pencil.

1 Here are the names of three life processes. The letters
have been jumbled up. Can you sort them out?

t g h o r w   _____

t o n n i i u r t   _____

t e m o m e v n   _____

Activity 2 |22|

You will need: a pen or pencil.

 1 Look at the pictures and sort the things into two groups:

a living things

b non-living things.

Activity 2 *(continued)* 22 – 23

2 Write down your two lists:

Living things		**Non-living things**	
_____	_____	_____	_____
_____	_____	_____	_____
_____	_____	_____	_____
_____	_____	_____	_____
_____	_____	_____	_____
_____	_____	_____	_____

3 For each of the two groups of things, write at least three sentences that describe their common features (characteristics).

Group 1

All the living things _____

All the living things _____

All the living things _____

Group 2

The non-living things cannot _____

The non-living things _____

The non-living things _____

4 Go outside and collect at least four examples of each group.

5 Back in the classroom, arrange your two groups for the class to see.

6 Look at the groups made by others in the class.

Healthy diets and exercise

There are different ways of grouping foods and this picture shows
food of six types.

1   2   3   4

5   6   7   8

9   10   11   12

13   14   15   16

Activity 4: What makes a meal healthy or unhealthy? ⌊27⌋

You will need: a pen or pencil.

1 Look at the food types shown in the picture.

Match the names of each food to the groups listed below:

| fruits   fats and oils   vegetables   staple foods |
| foods from animals   legumes |

1 _____   9 _____

2 _____   10_____

3 _____   11 _____

4 _____   12 _____

5 _____   13 _____

6 _____   14 _____

7 _____   15 _____

8 _____   16 _____

Activity 4: What makes a meal healthy or unhealthy? (continued) 📖 28

 2 Choose foods to plan a meal that is *healthy*.

Draw a picture of the meal.

Write the name of the foods and their food group here:

Food   **Group**

_____   _____

_____   _____

_____   _____

_____   _____

Activity 4: What makes a meal healthy or unhealthy? (continued) 28

 3 Choose foods to plan an *unhealthy* meal.

Draw a picture of the meal.

Write the name of the foods and their food group here:

Food   **Group**

_____   _____

_____   _____

_____   _____

_____   _____

4 What food groups did you use? Display your drawings.

5 Explain why the meals are healthy or unhealthy.

Activity 5: Why do people exercise? 30

You will need: a pen or pencil.

1 Collect as many sources of information about exercise and people who are athletes and sports players as you can. List them here. Cut and stick pictures and headlines here from magazines or newspapers too.

2 Use the materials to make notes about *why* people exercise.

_____

_____

_____

_____

_____

Activity 5: Why do people exercise? *(continued)* 30

3 Discuss with your group the advantages of taking exercise.

4 Record your group's ideas in the table. The first row has been filled in for you.

Advantages of exercising	Disadvantages of not exercising
The heart is kept healthy	Muscles are weakened

5 Share the group's ideas with the class.

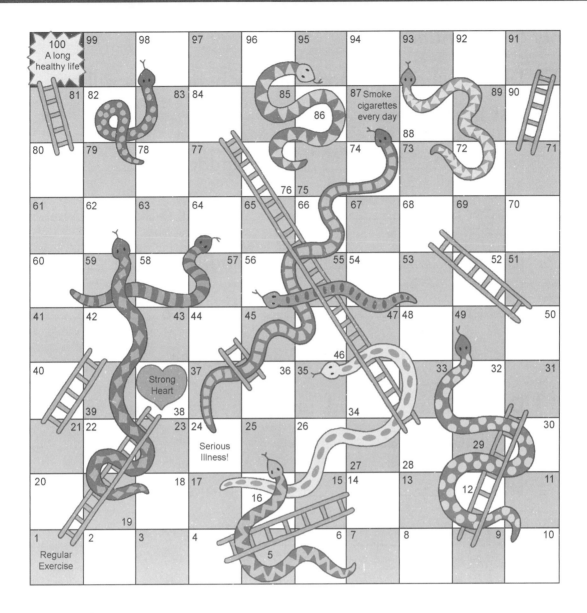

Activity 6 |32|

You will need: card or paper, a ruler, coloured pencils and a pen or pencil.

1 Use the game shown here, or copy it onto another piece of card or paper. Draw the squares first. Then put in the ladders and snakes.

Activity 6 (continued) 📖 32

2 Decide what the good, healthy behaviours will be.
List them here:

ladder 1 _____   ladder 5 _____

ladder 2 _____   ladder 6 _____

ladder 3 _____   ladder 7 _____

ladder 4 _____   ladder 8 _____

Put one at the *bottom* of each ladder.

3 Decide what the bad, unhealthy behaviours will be.
List them here:

snake 1 _____   snake 6 _____

snake 2 _____   snake 7 _____

snake 3 _____   snake 8 _____

snake 4 _____   snake 9 _____

snake 5 _____   snake 10 _____

Put one at the *top* of each snake.

4 Colour in the snakes and ladders to make the
game more interesting and fun.

5 Play the game with your group.

Foods that damage health

Activity 7: Investigate the nutrition information on packaged foods ⎸34⎸

You will need: food packaging and a pen or pencil.

1 **Collect packaging from a variety of foods.**

a Look at the nutrition information on the packaging.

b Use the information to sort the items into four groups:
  i low in fat   **iii** high in fat
  ii low in sugar **iv** high in sugar

2 **Display your groups of packaging.**

a Discuss your groups with the people you are working with.

b Record what you have found.

Draw and name each type of the packaging under the correct group heading. Some foods will belong to more than one group.

Low in fat	Low in sugar
High in fat	**High in sugar**

3 **Share your findings with the class and look at what your other classmates have found.**

Activity C

You will need: a pen or pencil.

1 Keep a food diary over a week.

Use the table below to record what you eat each day: at home, at school, or when you have meals or snacks in other places.

2 Put a tick against the foods you eat each day. For example, if you eat an orange on Sunday, tick the 'Fruit' row under 'Sun'.

Foods	Sun	Mon	Tues	Wed	Thur	Fri	Sat
Fruit							
Nuts							
Vegetables							
Rice							
Bread							
Pasta							
Milk							
Cheese							
Meat							
Eggs							
Fish							
Sugar							

Foods	Sun	Mon	Tues	Wed	Thur	Fri	Sat
Butter							
Fat spread							
Chips							
Cakes							
Chocolate							
Biscuits							
Fruit juice							
Fizzy drink							
Sweets							

3 **At the end of the week, answer these questions:**

a Which food did you eat **most** often?

_____

b Did you eat the same foods on any **two** days in the week?

_____

If you did, which days were they?

_____

c Which food did you eat **least** often?

_____

Activity D

You will need: a pen or pencil.

Complete these sentences using these words (you may need to use some words more than once):

> varied   nutrition   living   life   drink   safe
> movement   non   fat   diet   world   sweets   eat
> senses   fatty   damage   fried   healthy

1 Growth, __ __ __ __ __ __ __ __, reproduction and

__ __ __ __ __ __ __ __ __ __ are all __ __ __ __ processes.

2 Only __ __ __ __ __ __ things do these things. This is

how we know if things are living or __ __ __-living.

3 All the things we __ __ __ and __ __ __ __ __ are called

our __ __ __ __. It should be __ __ __ __ __ __ to keep

our bodies __ __ __ __ __ __ __.

4 We can __ __ __ __ __ __ our health if we eat lots of

sweet or very __ __ __ __ __ foods, such as sticky

__ __ __ __ __ __ and things __ __ __ __ __ in oil

or __ __ __.

5 Our __ __ __ __ __ __ help to keep us __ __ __ __ by

telling us about the __ __ __ __ __ around us.

Activity 8: Investigate the senses 36 – 37

> **You will need:** three different foods, three different objects, a bag or a cloth, three plastic pots or jars, elastic bands, three materials that smell and a pen or pencil.

1 Prepare tests for the other groups in the class. To find the answers to each test, they will have to use one of their sense organs.

One test is for the nose.
One test is for the tongue.
One test is for the skin.

2 Test A: Put the objects inside the bag, or on a desk covered by the cloth.

a Which test does your group think this one is?

b Discuss with your group.

3 Test B: Keep the foods separate and cut them up into small pieces, on pieces of paper.

Use the blindfold to cover the eyes of learners doing this test.

a Which test does your group think this one is?

_____

b Discuss with your group.

Activity 8: Investigate the senses *(continued)* |37|

4 Test C: Put a small amount of a material in a pot or jar and cover it with a paper lid. Keep the paper in place with an elastic band. Make small holes in the paper with your pencil point. If the jar is transparent, cover it completely with paper so that the contents cannot be seen. Do the same for each of the three materials you have. Label the pots 1, 2 and 3.

a Which test does your group think this one is?

_____

 b Tell your group what you think.

Activity 8: Investigate the senses *(continued)* 38

Your group has just completed tests A, B and C.
Now complete the following questions.

 5 Invite learners from other groups to try your tests.

    a In each test, they have to use *one* sense to identify the objects, foods and materials.

     b Keep a record of their answers for each of the tests.

    c Compare the results and find the easiest parts and hardest parts of your tests.

6 Visit another group and do their tests.

 7 Record your answers in the table below.

Test	Object/material	Organ	Sense
A			
A			
A			
B			
B			
B			
C			
C			
C			

Activity 8: Investigate the senses *(continued)* |38|

8 Compare your results with others in the group.

   a Which was the hardest thing to identify?

_____

   Why?

_____

   b Which was the easiest to identify?

_____

   Why?

_____

(a) (b) (c) (d) (e) (f) (g) (h) (i) (j) (k) (l) (m) (n)

Activity 10 📖43

You will need: a pen or pencil.

1 Look at these pictures of living things.

Activity 10 *(continued)* 43

a Discuss with the people you are working with how you will sort them into groups.

Answer: _____

_____

b Record your groups using the letters **a** to **n** to identify the members of the groups.

c Choose a name for each of the groups.

There are boxes below for up to seven groups, but you might decide that there are fewer than seven or more than seven.

Name: _____	Name: _____
Name: _____	Name: _____
Name: _____	Name: _____
Name: _____	

2 Discuss how you will explain the way you sort and name the groups.

Answer: _____

 **3** **Share your results with the class.**

a Explain what you have done.

b Listen to the answers from your classmates and ask them for explanations.

---

43 Here are the jumbled names of some groups of living things for you to sort.

e s r t p e i l _____

s b e h r _____

b r h u s s _____

l a a m m m s _____

s i f h _____

e r t s e _____

i s b r d _____

Activity E

You will need: a pen or pencil.

Complete these sentences using these words (you may need to use some words more than once):

> warm   groups   birds   mammals   plants
> scaly   trees   fish   classify   birds   animals
> herbs   reptiles   visible

1 We can __ __ __ __ __ __ __ __ (sort) living things into

several __ __ __ __ __ __. The two main ones are the

__ __ __ __ __ __ and the __ __ __ __ __ __ __.

2 Fish, __ __ __ __ __, mammals and __ __ __ __ __ __ __ __

are some of the groups of animals.

3 Plants have groups called __ __ __ __ __, shrubs and

__ __ __ __ __.

4 The skin of animals is one __ __ __ __ __ __ __ feature

used in sorting them. It can be hairy (the

__ __ __ __ __ __ __) or __ __ __ __ __ (the reptiles and

the __ __ __ __).

5 The only animals with feathers are __ __ __ __ __.

They are like __ __ __ __ __ __ __ in one way – they are

all __ __ __ __ blooded.

Exploring properties

Activity 1: Find out about the properties of materials 47

You will need: objects from your classroom and a pen or pencil.

1 Walk around the classroom and find five objects made of *different* materials.

2 Take them to your desk and observe each one carefully.

   a Look at them, feel them, smell them.

   b Think of words to describe the properties of the materials.

3 Record the names of the materials and the properties you have found. Use this space to record your results.

4 Display your materials and your record for the class to see. Look at what other learners have found.

You can also sort non-living materials into groups using their properties.

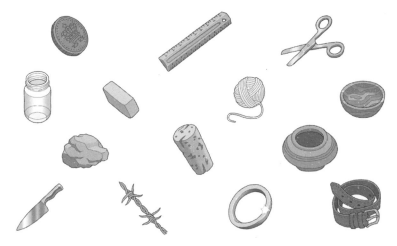

Activity 2: Which materials have common properties? 49

You will need: materials from in and around your classroom, and a pen or pencil.

 1 Look at the objects shown in the pictures.

a Sort them into groups based on the *properties* of their *materials*.

b Record your groups here:

Property: _____

Property: _____

Property: _____

Activity 2: Which materials have common properties? *(continued)* 50

Property: _____

Property: _____

Property: _____

2 **Share your groups with the class.**

a Explain why you have sorted the objects that way. What characteristics (properties) did you look for when sorting the objects?

_____

_____

 b Ask others to explain their groups.

Activity 2: Which materials have common properties? (continued) 50

3 Move around the classroom and outside looking for different materials.

a Draw what you find and write down their names.

b Share your findings with the class. Make a display for the others to look at.

4 What characteristics do the different materials have?

_____

_____

Tell the class what you think.

Activity 3: What materials are magnetic? ⌊51⌋–⌊52⌋

You will need: a pen or pencil.

1 In school, plan with your group how you will find magnetic materials in your classroom and outside.

2 Write down your plan.

  a Decide how you will do a fair test and how you will record your results.

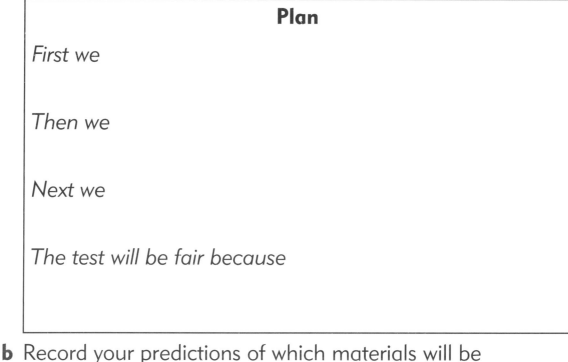

Plan

First we

Then we

Next we

The test will be fair because

  b Record your predictions of which materials will be magnetic and which will not.

I predict that:

_____ *will be magnetic because* _____

_____ *will not be magnetic because* _____.

3 Show your plan and predictions to your teacher.

Activity 3: What materials are magnetic? *(continued)* 52

4 Collect the things you need for your test. List them here:

a Test at least four different materials in the room and four others outside.

b Record the results each time. Ask for help if you need it.

You can make notes here:

5 Look at your results. If you find any patterns, try to explain them.

a Compare your results with your predictions. Try to explain why your predictions were not always correct.

b Come to a conclusion about magnetic materials and write it down.

6 Share your results, conclusion and explanation with the class.

7 Listen to the other groups and try to make a generalisation about the class results.

Activity 4: Identify which materials are magnetic 53 – 54

You will need: a pen or pencil.

1 a 'copper' coin   2 an iron nail   3 a drinking glass   4 a wooden spoon

5 a steel spoon   6 a woollen hat   7 a cotton sock   8 a plastic bag

9 a ball of string   10 a stone   11 a shell   12 a ceramic plate

13 A gold ring   14 a metal gate   15 an iron cooking pot   16 a metal mug

1 Here are some pictures of some objects. Read their names and think about what you have found out about magnetic materials.

 2 Choose the objects you are *sure* are magnetic. Write down their names and numbers.

Activity 4: Identify which materials are magnetic 53 – 54

3 Choose the objects you are *sure* are not magnetic. Write down their names and numbers.

4 Make a list of all the objects left that you are not sure about.

Can you explain why you are not sure about them?

_____

_____

5 Identify eight metal objects in the classroom.

a Now test them to find out which ones are magnetic metals.

b What do you find?

_____

_____

Activity 5 |55|

You will need: a pen or pencil.

1 Think of examples of how we use metals.

  a Write down some of the uses of metal.

  _____

  _____

  _____

  b Can you think of the names of the metals used in your examples? Write them down too.

  _____

  _____

2 Make drawings of three of the examples you have chosen.

Example 1:	Example 2:	Example 3:

3 Display your list and drawings for the class to see. Now have a look at what other learners have done.

Activity 6: Think about which properties are useful in materials and why 58 – 59

You will need: a pen or pencil.

1 In class, look at the pictures showing five materials used in many different ways.

a Discuss with your group what you think the five materials are.

b Write them down at the top of the five columns in the table below.

Materials and their uses

Uses	Material 1	Material 2	Material 3	Material 4	Material 5
	_____	_____	_____	_____	_____
1					
2					
3					
4					
5					

2 List all the uses of the materials you can see in each of the five columns.

3 Share your lists with the class. Add to your lists if you missed anything out.

Activity 6: Think about which properties are useful in materials and why *(continued)* 58 – 59

4 Take one example from each list and try to explain why the material is used in that way.

For example:

The boat is made of __ __ __ __ *because* _____.

_____ *wires are covered in* __ __ __ __ __ __ __

because _____.

_____

_____

_____

_____

_____

5 Share your sentences with the class.

6 Here are some terms describing some physical properties of the materials:

> transparent   insulating   waterproof   strong
> mouldable   hard   easily shaped   flexible
> lightweight   smooth

a On the following page, copy the words one under the other to make a list.

b Beside each word write the names of the materials that you think the words describe.

Activity 6: Think about which properties are useful in materials and why *(continued)* 58 – 59

Properties	Materials
_____	_____
_____	_____
_____	_____
_____	_____
_____	_____
_____	_____
_____	_____
_____	_____
_____	_____

c Share your lists with the class.

Activity F

You will need: a collection of objects and a pen or pencil.

1 Collect items at home that match these properties.

> rough   shiny   soft   elastic   hard   smooth

Collect one item for each property.

2 Draw pictures of the items and write their names in the spaces below.

Rough _____

Shiny _____

Elastic _____

Hard _____

Smooth _____

Soft _____

Activity G

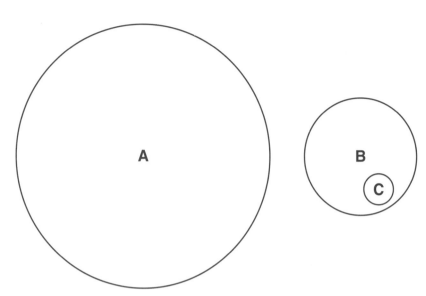

You will need: a pen or pencil.

1 Look at the diagram of the three circles.
They stand for three groups of materials:

> **magnetic metals**   **non-metals**   **metals**

2 Match the circles to the three groups of
materials and write the answers below:

A is the group of _____

B is the group of _____

C is the group of _____

3 Write down two examples of each group:

Group A:_____ and _____

Group B: _____ and _____

Group C: _____ and _____

Chapter 4: Forces and motion

Push and pull

A force is either a push or a pull. Sometimes both types of force act together.

(a)

(b)

(c)

(d)

(e)

(f)

(g)

(h)

(i)

(j)

(k)

Activity 1 [61]

You will need: a pen or pencil.

 1 Look at the situations shown in the pictures on page 55. Sort them into three groups:

a those that show a push

b those that show a pull

c those that show both a push and a pull.

 2 Write down the three groups using the letters beside the pictures.

a those that show a push:

_____

_____

b those that show a pull:

_____

_____

c those that show both a push and a pull:

_____

_____

3 Share your groups with the class.

Activity 2: How does a force meter work? |64|

You will need: a force meter, and a pen or pencil.

1 Collect some objects that your group wants to use for measuring force.

2 Can you work out how a force meter works? Have a go.

3 Record your results in the table below. The first row has been filled in for you.

Table of results

Object names	Push object	Pull object	Force (N)
Stone	No	Yes	5

4 Measure the force you use to pull or to push each object that you have chosen to investigate.

a Record each time what force you used – a pull or a push.

b Record the force meter readings.

5 Share your results with the class. Make a display of all the results.

Activity 3: Exploring how forces can stop or start the movement of a rolling object 67

You will need: a flat, smooth surface to make a slope (for example, a desk top, table top or wooden plank), a ball or toy with wheels, paper and a pen or pencil.

1 Take a ball, toy with wheels or some other object that can roll down a slope.

2 Use a plank, desk or table top or other flat, smooth surface to make a slope.

3 Plan how your group will investigate *two* of the *effects of force* on the movement of your rolling object: *stopping and starting movements*.

Plan
First we
Then we
Next we
The test will be fair because

4 Share the group's plan with the teacher.

Activity 3: Exploring how forces can stop or start the movement of a rolling object *(continued)* 67 – 68

 5 Carry out your exploration of the effects of force on movement. Record the following three things each time:

   a your prediction of what will happen

   b what you did

   c what happened.

 6 If you fail to have the effect you planned and predicted, try another way until you are successful.

   Record your results in a table using the headings below:

Stopping movement

What we predicted	What we did	What happened	What worked well	What we would do differently next time

Starting movement

What we predicted	What we did	What happened	What worked well	What we would do differently next time

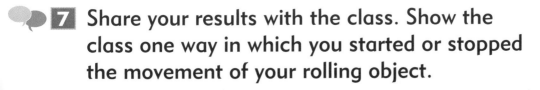 **7** Share your results with the class. Show the class one way in which you started or stopped the movement of your rolling object.

Changing shape

Activity 4: What happens to the shape of objects when you use force? |70|

You will need: objects to experiment on and a pen or pencil.

1 In class, plan your group's exploration of shape changing.

   a Choose at least two objects to explore. List them here:

   b Decide which type of force you will use for each one –
   a pull or a push, or both. Write 'push' or 'pull' next to
   each object name in the box above.

2 Make a small drawing of each object to show its shape
   before you tried to change its shape.

Objects before force was used

Activity 4: What happens to the shape of objects when you use force? *(continued)* |70|

 3 Record your predictions and results in the table below.

Object name	Force/s used	Prediction	Drawing of object after force used	Was prediction correct?

a Try to change the shape of each object.

b Record the result in a small drawing. It should show the shape *after* the force was used.

4 When all objects have been tested, compare your predictions with your results and fill in the last column with 'yes' or 'no' for each one.

Forces speed up, slow down and change the direction of movement

Activity 5: Explore the movement of objects on a slope 📖73

> **You will need:** a flat, smooth surface to make a slope (for example, a desk top, table top or wooden plank) and a pen or pencil.

1 In class, plan your exploration of speeding up and slowing down objects, using a slope.

   a Choose three different objects to test.

   b Write down your plan and show it to your teacher. You must include the question you are trying to answer, such as:

     'What happens if _____ *?'*

     or

     'How can we make _____ *?'*

2 Complete the table to record your results.

     Write your predictions down in the table before you start the exploration.

Object	Prediction	Result

Activity 5: Explore the movement of objects on a slope (continued) 73 – 74

3 Explore each object one at a time until you have an answer to your question.

a Record what happens as you use forces to speed up the objects.

b Record what happens as you use forces to slow down the objects.

4 Observe what happens when the object moves:

a off the board and onto the floor

b off the board and onto the table, or other surface.

5 Record your observations in the table.

6 Look at your results and discuss them with your group.

a Compare your results with your predictions.

_____

_____

_____

b Come to conclusions about how movements were speeded up or slowed down.

Movement speeds up when _____.

Movement slows down when _____.

Activity 6: Explore changing the direction of moving objects

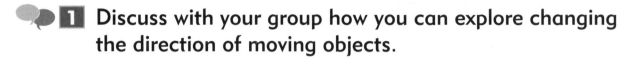

You will need: some objects to test and a pen or pencil.

1 Discuss with your group how you can explore changing the direction of moving objects.

2 Plan a fair test and show your plan to your teacher.

Plan

3 Make careful observations and record what you observe. Start by using the space here to draw, make tables or charts, or write notes. Use extra paper if you need to.

4 Look at your results and come to some conclusions about how direction of movement can be changed. For example:

Objects can be made to change direction by

_____

_____

5 Share your results and conclusions with the class.

Activity H

You will need: a pen or pencil.

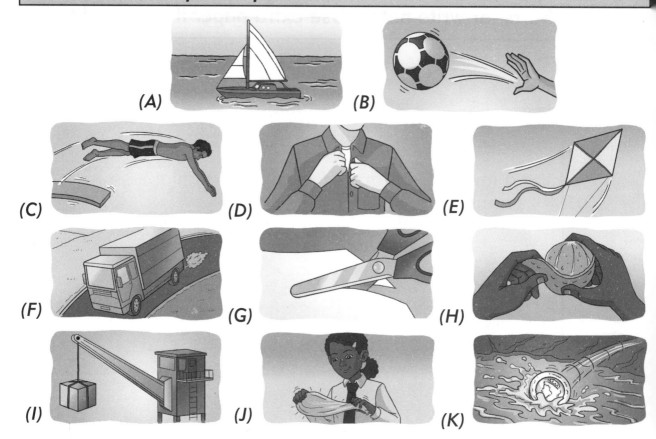

(A)

(B)

(C)

(D)

(E)

(F)

(G)

(H)

(I)

(J)

(K)

1 Look at the situations shown in the pictures and sort them into three groups:

a those that show a push

b those that show a pull

c those that show both a push and a pull.

2 Write down the three groups using the letters beside the pictures.

a Push: _____

b Pull: _____

c Push and pull: _____

Activity I

You will need: shoes with different sole types, a flat surface to create a slope, a ruler and a pen or pencil.

1 Collect at least two different shoes. Get the same size if possible. Try to find shoes with different types of soles. For example, shiny, smooth soles and rough or rubber soles.

2 Find a large book, or a tray, or a small plank of wood to use as a slope.

3 While the 'slope' is still flat on the floor, place a shoe on it. Make sure the back of the shoe is in line with the edge of the slope.

4 Slowly lift the slope. Observe what happens to the shoe. When it starts to move down the slope, stop lifting. Use the ruler to measure how far you lifted the book or tray or plank. Make a note of the measurement. Label it 'A'.

5 Predict how the second shoe will move down the slope.

Write your prediction here:

When the shoe has a shiny, smooth sole I predict that

_____

_____.

When the shoe has a rough or rubber sole I predict that

_____

_____.

Activity I *(continued)*

6 Now start the test again with the second shoe. When this shoe starts to move, use the ruler to measure how far you lifted the slope. Make a note of the measurement and label it 'B'.

7 Compare the two measurements and see if your prediction was correct or not. Write your result here:

_____

_____

8 If you have more shoes to test, use the same method for all of them.

9 When all the testing is finished, look at your results and try to explain them. Write your explanation here:

_____

_____

_____

_____

10 Draw pictures of the soles of two shoes that began to move at different levels of slope.

This shoe moved on the lower slope:

This shoe moved on the higher slope: